THE KEEPER OF TIME
PAINTINGS OF PALO DURO CANYON

by Len Slesick

AuthorHouse™
1663 Liberty Drive, Suite 200
Bloomington, IN 47403
www.authorhouse.com
Phone: 1-800-839-8640

©2008 Len Slesick. All rights reserved.

No part of this book may be reproduced, stored in a retrieval system, or transmitted by any means without the written permission of the author.

First published by AuthorHouse 5/15/2008

ISBN: 978-1-4343-8212-2 (sc)

Library of Congress Control Number: 2008903722

Printed in the United States of America
Bloomington, Indiana

This book is printed on acid-free paper.

TABLE OF CONTENTS

INTRODUCTION	2
ACKNOWLEDGMENTS	4
PALO DURO CANYON — KEEPER OF TIME	5
THE PEOPLES OF THE CANYON	6
THE PARK THAT ALMOST WASN'T	9
PAINTERS OF THE CANYON	13
60 PAINTINGS OF PALO DURO CANYON BY LEN SLESICK	15
CHRONOLOGY	75

INTRODUCTION

By Len Slesick

A gift was bestowed upon me when I ventured to the Texas Panhandle. Within a month of moving to Amarillo in 1965, I visited Palo Duro Canyon State Park for the first time. The paintings in this book are from many trips since then, walking not only the main canyon but also most of the tributaries from bottom to top. The rugged beauty of Palo Duro Canyon is ever changing; hazy days, the crisp feeling after a snow, the imagination of night. The subject matter may not be new, but the approach to the subject is different in many ways. It's my imagination at play, not merely pictures, but feelings — some extremely visual, others allowing the mind to flow with that bit of inner self that finds its way into a painting.

I was asked why I paint the canyon. And I had no answer. After giving this considerable thought, I came up with the word "Curiosity."

When I began exploring the canyon, curiosity was a driving force. What is around the next curve? What is up the next canyon? Something is beyond the next hill — go and find it. At times there was nothing outstanding, but at other times I would be greeted with a spectacular sight or view.

I began painting during the early 1970s, and after a few years, when some proficiency began to sink into my head, I remembered some of the locations I had ventured to in the canyon. So off I went, to paint and photograph. Soon I was finding new sights to paint, maybe not that I had missed seeing before, but now I was looking at things in a different way. There is something to paint everywhere you look. Not only the grand panoramic vistas, but also little features such as a rock slide creating wonderful shapes or a group of prickly pear cactus arranged in an interesting pattern. From more than 100 paintings of the canyon, I have chosen 60 to be in this book. Some are quick studies, on location, others from photographs taken on my treks, and the contemporary pieces, straight out of my feelings. Many of the latter have taken on the "Three Mesa" subject, repeating this simple theme in numerous attitudes.

My favorite palette is simple but well thought out. I begin with the complementary colors, cobalt blue and cadmium orange, and their analogous colors, blue violet (ultramarine violet), blue-green (viridian), red orange (cadmium red light), and yellow orange (cadmium yellow deep). Most of the time I will also add cadmium yellow light for a total of seven paints and white. I like to use soft titanium for the initial mixing and flake white during the painting. To control intensity I'll mix about five parts blue to one part orange for a neutral gray. This will also create my greens. I avoid earth colors with this palette.

I enjoy painting in the comfort of a studio but also like to paint outdoors.

Painting on location is best in the early morning or late afternoon when shadows are long, but the abundant sunshine causes light to change in a matter of minutes. My feeling toward plein air painting is to do a brief study, maybe 20 to 30 minutes, being certain to catch the shape, color, and value in the shadowed areas. These will change quickly and will not be true in a photograph. (Darks are always too dark in a photo, and many times little or no color shows.) The shape of the sunlit part of your painting will move with the sun, but you have a little more time to catch the colors and values before significant modification takes place. I usually take a photograph after my initial layout, but before starting colors and values. I can then finish work back at the studio; maybe not pure plein air, but it works for me. A number of these short studies can be done in one day, and in Palo Duro Canyon, you merely change position to see another spectacular view or a detail to paint a vignette. When skies are overcast you can spend more time on a single location since the light source is not changing as quickly.

ACKNOWLEDGMENTS

I would like to express thanks to Dr. Fred Rathjen and Dr Gary Nall, both retired from the history department of West Texas A & M University. Their assistance in editing the initial manuscript has kept me historically and grammatically correct. To Michael Grauer, curator of art and associate director of territorial affairs at the Panhandle Plains Historical Museum in Canyon, Texas, for his knowledge and guidance of the painters of the canyon, to David Corbin for photographic expertise, and finally to my wife Ginny for 50 years of guidance and tolerance.

PALO DURO CANYON — KEEPER OF TIME

After traveling over a vast expanse of grassland, the sight of Palo Duro Canyon is still as awesome today as it must have been when the Spaniards first came to its edge almost 500 ago. Flat, steppe-like land continues level and then very suddenly opens to a chasm with walls from six to 800 feet high; multi-colors, each telling its own story of a different age: red-oranges, violets, browns, yellows, and grays changing not only by layers but by light, time of day, and season of the year. Clouds passing overhead mute the colors only to have them burst into flames again with the reappearing sunlight.

But to account for the history and time line of the canyon, we do not start at the top, but rather at the bottom. More than 200 million years ago an arm of the Permian Sea covered this part of the world. The alternating rise and fall of sea level deposited the colorful red-orange beds making up the lowest portion of canyon walls and formations. Intermittent dry spells or the receding of the sea left thin white lines of gypsum, the sea returns — more red deposits — and retreats — more white, back and forth over tens of thousands of years until a time record builds upward. A final, gradual southwest retreat of the Permian Sea brought this period of time and history to a close.

A new time, a new era, and more colors are laid on top of the red beds, the yellows, violets, and grays of the Tecovas formation. These, along with the earlier red deposits, form the dancing "Spanish Skirts." Rather than being deposited at the bottom of a sea, as were the red beds, these layers washed in from the southeast, similar to the formation of the Mississippi Delta below New Orleans. We move higher. Massive layers of sandstone form a number of ledges and the upper steep slopes of the Trujillo formation. Finally, we have reached the top of the canyon, the caprock, a hard layer of caliche, much more resistant to erosion than the underlying layers. This, then, was the building upward of the canyon walls, colorful levels hidden under the ground and protected by the caprock. But slowly at first, and then more rapidly, erosion began its work by causing breaks to appear in the hard top layers. Mostly water and frost action forced cracks to widen and open, exposing the softer layers beneath. Now a rapid breakaway takes place, and in less than a million years the colorful walls of the canyon and its tributaries are exposed. Time is laid out before the viewer. With each new rain, the process continues, ever changing the shape of Palo Duro Canyon.

THE PEOPLES OF THE CANYON

After speaking of ages in the tens of millions of years to develop the structure of Palo Duro Canyon, the time line of man seems infinitesimal. Ten to twelve thousand years ago the big game hunters of mammoth and giant bison wandered the plains, with some evidence pointing to their existence in the canyons. The climate was wet and cool after the last ice age. Slowly, a warmer and drier climate of the altithermal set in. A 3,000-year drought of hot and dry weather established itself. The large animals abandoned the area, and the people who hunted them left. A few small bands, similar to the nomads wandering today's world deserts, may have remained. Then a more moderate climate returned about 2000 BC, and animal life returned, but these were similar to the animal life we find today. This in turn caused man to return, this time not only as hunters, but also as gatherers, and eventually these people turned to farming.

Pictographs in a Shelter
Palo Duro Canyon State Park

Mortar Hole
Palo Duro Canyon State Park

The first Europeans to view the canyon were the Spaniards of the sixteenth century. Tales telling of the gold of "Eldorado" had them leave Mexico, travel into today's southwest, and cross the south plains searching for the fabled but mythological city of gold. Descriptions of their sighting the canyons held the Spaniards in awe. These were probably the first historic people to visit Palo Duro Canyon and may have coined the name "Palo Duro" from the Spanish for hard wood, referring to the junipers that thrive in the canyon. Although various other European groups sighted Palo Duro Canyon during the eighteenth and nineteenth centuries, it was not until 1852 that the first Anglo-Americans left a record of their admiring the panoramic view. Captain Randolph Marcy was ordered to "…make an examination of the Red River and the country bordering upon it..." Marcy reached the headwaters of the Red River and the main branch called "Ke-che-a-qui-ho-no," Comanche for Prairie Dog Town River. They were as overwhelmed then as people are today when visiting the canyon for the first time. "We all with one accord stopped and gazed with wonder and admiration upon a panorama which was now for the first time exhibited to the eyes of civilized man. Occasional might be seen a good representation of the towering walls of a castle of the feudal ages, with its giddy battlements pierced with loopholes, and its projecting watchtowers standing out".

Next came the buffalo hunters of the mid 1870s. Kansas, the center of the buffalo hunt, was depleted of most animals, and the territory of the Texas Panhandle looked very inviting — dangerous but inviting. Hunters worked the plains of the Panhandle and eventually Palo Duro Canyon, until the Battle of Adobe Walls near the Canadian River opened the Red River Indian War. A decisive battle took place at the confluence of Cita Canyon and the Prairie Dog Town Fork of the Red River. This is just south of the confines of the state park. On September 27, 1874, Colonel Ranald Mackenzie had his troops descend a narrow trail to the canyon's bottom. The attack ensued with a number of running battles, including one upstream, which ended inside the limits of what is now Palo Duro Canyon State Park. Mackenzie proceeded to destroy the camp and supplies and captured 1,448 horses. Tonkawa guides kept 400, while the remainder was destroyed to ensure that the Indians would return to the reservation at Fort Sill, Oklahoma territory.

The first Anglo to settle in the canyon was Charles Goodnight. Goodnight was a Texas Ranger during the Civil War and while in the region observed the rugged edge of the canyon with its high walls for containment of cattle and protection from winter winds and cold. Water also seemed in evidence. With the close of the Indian hostilities in the Panhandle, Charles Goodnight, then ranching in Colorado, drove a herd of cattle south and settled in Palo Duro Canyon, becoming the first rancher in the Texas Panhandle. This was during the year 1876.

THE PARK THAT ALMOST WASN'T

Palo Duro Canyon emerges from the seemingly flat, endless landscape of the plains, and the people of the region always knew it was something special. The turn of the twentieth century found a meager population on the plains of the Panhandle that was aware of the treasure they had in their own backyard. As early as December 1906, the Canyon City Commercial Club passed a resolution asking for the establishment of a national park on the upper end of Palo Duro Canyon. Two years later a bill was introduced in Congress for the purchase of a park in the state of Texas to be known as "The Palo Duro Canyon National Forest Reserve and Park." The bill died and was reintroduced in 1911 and again in 1915. By 1917 it was apparent that a bill was unlikely to pass in the near future. Yes, the treasure was here, but few others knew about it. Nothing more was done on a national level, but on May 17, 1931, the "Grand Opening" of Palo Duro Park took place, but this was still on private land. Finally, the Texas legislature approved use of anticipated park revenues to repay the loan for purchase of the property, the act was signed May 29, 1933, by then-governor Miriam A. Ferguson. So the Panhandle could have a park — if we paid for it.

About this same time, the Great Depression had settled on the United States, and the National Park Service saw Palo Duro Park as a likely site for Civil Conservation Corps projects; however, there was some concern as to the financing of the park. Only after the Texas Park Board agreed to service the debts through local revenues and many of the Panhandle's leading citizens signed a petition stating the approval of the purchase and repayment scheme did the Texas Relief Commission finally give their authorization to approve CCC projects in Palo Duro Park on June 29, 1933. Immediately, the army built an encampment, brought in water and a phone line, and with generators supplied electricity. The first CCC workers arrived July 12, 1933. By November 26, the road to the bottom of the canyon was completed and officially opened. Palo Duro Canyon State Park had its official opening July 4, 1934.

Souvenir from Park Opening 1934

The Depression continued. Vast dust storms raked the region, and attendance at the park fell well short of meeting the financial obligation. World War II added to the problem with gasoline rationing cutting travel and tourism. After the war, a foreclosure on the park loan appeared imminent. With little help coming from Austin, once again Panhandle residents realized that the park operation would have to be their responsibility. By the late 1940s, what had seemed an impossible task began to turn. Recreation and travel increased the park's income, and in 1949 Amarilloan John McCarty took over the operation of the park concessions. The promoter brought buffalo, deer, and longhorns to the park and held a "treasure hunt" for two years. Find a coin dropped from an airplane and win an attractive prize: a new car, a trip to Rio de Janeiro, or a horse. Park revenues quadrupled in just 10 years, sorely needed improvements were made, and all financial obligations were paid during the period 1960 through 1966. Palo Duro Canyon State Park had survived.

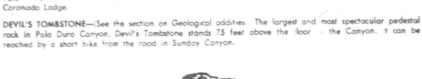

Early Entrance Pamphlet
Palo Duro Canyon State Park

Treasure Hunt Coins
1949 and 1950

Panhandle residents were proud of their work, but there was resentment for the lack of "downstate" assistance. Senator Grady Hazlewood stated, "I was a member of the Senate when the Legislature voted one and a half million dollars to buy Big Bend State Park. Session after session I saw Palo Duro State Park get $250.00 for maintenance, although it contained 15,000 acres, while other state parks received anywhere up to $1,500.00 and some more. I introduced an appropriation bill to buy Palo Duro State Park but with no success. Finally the legislature very 'generously' agreed to let us have this park if we would pay for it ourselves. I know of no other state park in Texas that has been paid for by the people of the local area... When paid for by our local efforts it will belong to all of Texas — not us." An addition to the park during 2002 was the acquisition of the Canyoncita Ranch, 2,036 acres taking in a large portion of North Cita Canyon. A more recent acquisition of 7,600 acres from the Harrell Ranch increased the park total to 26,275 acres.

During the summer of 1966, a lone horseman appeared on the edge of the canyon holding the Texas Flag. The musical drama *Texas* was born. Nationally acclaimed, the play has entertained audiences in the millions from around the world.

A MUSICAL ROMANCE OF PANHANDLE HISTORY

1966 Program
Paul Green's play, TEXAS

PAINTERS OF THE CANYON

The sheer magnitude and color of Palo Duro Canyon has attracted artists for more than a century. Frank Reaugh, born in Jacksonville, Illinois, came to Texas in 1876 and was one of the earliest painters of the canyon. Reaugh would take students to Tule Canyon, feeling they were not ready for the vivid colors of Palo Duro; it was too much for them to absorb at one time. Another school setting was Isabel Robinson's Palo Duro School. Isabel was the head of the art department at West Texas State Teacher's College and held classes in the canyon 1936–1943. Harold Bugbee assisted as an instructor for a part of this time.

The best known of the canyon painters was Georgia O'Keeffe. Her free spirit took immediately to the vast sky and the openness of the prairie. "The feeling of being a small dot in space standing on the West Texas Plains. Grass opens to a vast canyon ready to devour you." Georgia O'Keeffe taught in the Amarillo public school system in 1912–1913, but it was during her tour as a teacher at West Texas Normal College in 1916–1918, that O'Keeffe painted the Panhandle landscape and Palo Duro Canyon. In 1972 Georgia was quoted as saying, "I lived on the Plains of North Texas for four years. It is the only place I have ever felt that I belonged — that I really felt at home, that was my country." She was by no means the only painter to be inspired by the canyon. Victor Higgins of the Taos Art Colony painted in Palo Duro Canyon during 1923–1924, and in 1926 Alexander Hogue spent 10 days sketching the canyon. Hogue studied under Frank Reaugh on three summer trips and during the 1930s was part of the Texas Regionalist movement. Wilson Hurley of the contemporary school of painters was known for his panoramic scenes of Palo Duro. More recently, murals by Ben Carlton Mead and Harold Bugbee grace the walls of the Panhandle Plains Historical Museum in Canyon, Texas, depicting some Palo Duro scenes in historical settings. The canyon's magnet continues to draw painters to its awesome beauty.

60 PAINTINGS OF PALO DURO CANYON BY LEN SLESICK

Sizes of paintings are in inches, with vertical measurements first.
Unless indicated, paintings are owned by Slesick Gallery

1. The Keeper of Time — 1998 — Oil on Canvas 24 x 36
— Collection of State National Bank of Groom, Amarillo, TX

How proud this sentinel stands, looking over the vastness of the park, displaying its years of knowledge and history as if to say, "I am the keeper of time". The Prairie Dog Town Fork of the Red River meanders through the bottom of the canyon. This lazy stream is responsible for carving the giant gorge from the flat tabletop of the Llano Estacado. Millions of years of time are exposed, each time frame a different color, a different piece in the puzzle of pre-history. One of my favorite painting locations is the first water crossing. Early morning light illuminates the west wall, offering hours of painting. By the afternoon the "Spanish Skirts" to the east are in full light. To the right, shadows and light compete with each other but never completely hide the features of a distant canyon. Catch the distance with muted colors.

2. The Lighthouse — 2000 — Oil on Canvas 28 ½ x 35 1/4

Painted, photographed, and described more than any other feature, the Lighthouse has become the trademark of Palo Duro Canyon. Freed to stand alone, this monument rises nearly as tall as the surrounding terrain from which it escaped. All the colorful layers found in the nearby walls are banded together, some reflecting the sunlight while others are muted in shadow. The cool blue-green sky was chosen as a complement and contrast to the warm red-orange terrain.

3. Dry Wash With Yuccas — 2003 — Pastel on Paper 10 x 14

Sometimes small features make an interesting painting. In this case, the curved shapes of the sides forming the dry wash caught my attention but were not enough to be the entire painting. I added the bold colors of the turquoise sky and the violet hills but left them flat and featureless to make the sand wash and yuccas more important. The sand forms graceful catenary curves in opposition to the yuccas perched atop the embankment like conquistadors ready to defend themselves with sword-like leaves. Opposition makes a painting work, whether it is color, shapes, or direction, in this case all three.

4. Rock Tumble — 2002 — Pastel on Paper 14 x 10 — Private Collection

What is so interesting about a pile of rocks? So many times I have walked by this scenario with total disinterest, but one day the great directional pulls struck me and a new painting was on the way. This deep, narrow passage with rock slides shows the effect gravity plays in shaping the canyon. Rains melt soft layers in the walls while the harder rock above fractures and tumbles to build interesting patterns. Look for shapes and patterns. They are the backbone of this painting. A strong diagonal at the top is countered by an opposing diagonal formed by rocks near the lower portion of the work. These are more important than color, even though I smuggled in another turquoise sky. It's not very large or apparent, but there it is.

5. Shapes in the Canyon — 1994 — Oil on Canvas 24 x 30

You hear about "Artistic Liberties." Well, this is a perfect example. It's not that I'm dissatisfied with nature — she led me to this spot — but now it's my turn to lead the viewer through this painting. All directional lines are placed to merge at a central vanishing point. Follow the streambed. Both banks move into the center and disappear. The light-tipped grasses sway inward on each side of the creek, helping to lead your eye toward the center. The shapes of the canyon walls in the background pivot from left to right around a central point. "Artistic Liberties": Just one more tool in the painter's bag of tricks.

6. Canyon Walls — 1997 — Oil on Canvas 48 x 60 — Collection of Perry Williams, Inc., Amarillo, TX

Shapes and angles are everywhere. These must be captured correctly in my initial layout if I expect the massiveness of the canyon walls to be of an imposing scale. The harder material near the top will form an abrupt drop, the softer lying as a slope, and finally the shear cliff to the bottom. The sky has been altered to form a shape. Rarely do I leave the sky as it looks since it is a great vehicle to tie the painting together with color or shapes or movement.

7. Red Canyon — 2002 — Oil on Canvas 14 x 18

The lowering sun of autumn brings a lazy feeling in the air. Winter is not far away, but for now it's mild and a fine day to paint. We are looking into the canyon from a viewpoint at the top. Reds are everywhere, and my palette will exaggerate this. Even the sky will be tinted with red, producing a sharp contrast to the deep greens of the cedars and the vivid yellows of cottonwoods as they wander aimlessly along the course of a tributary to the main stream.

8. Devil's Slide — 2004 — Oil on Canvas 16 x 20 —
Collection of Janette Dickerson, Amarillo, TX

The perfect time of day for painting is about an hour after sunrise when shadows are long and sunlit areas brilliant. The Devil's Slide is a knife-like structure catching the full effect of the sun on one side while creating deep shadows on the other. Paint the colors in the shadows first since they will be the first to change, then the warm sunlit face. I usually take a photograph as I start to paint for any finishing work at the studio. The photograph usually misses the great colors in the shadows, so I work carefully in that region while on location. Don't ignore the foreground. By use of dark areas and light, a tedious large part of the painting can become interesting and help hold space.

9. Fortress Cliff — 2002 — Pastel on Paper 8 x 14

A familiar landmark in Palo Duro Canyon State Park is Fortress Cliff, which forms a portion of the east wall of the canyon. I always envision a line of Comanche on horseback suddenly appearing along the edge, just the way we see them in the "western movies." Even though the cliff is the main topic, a large foreground cannot be ignored. To make it interesting, I've used slight changes in color and value. I do not want this to compete with Fortress Cliff, so a reverse of the rules takes place: small changes in the foreground and the brightest colors and largest value changes in the background.

10. Capitol Peak — 2001 — Oil on Canvas 8 x 10
— Collection of Eva Challis, Dallas, TX

As the sun rises over Palo Duro Canyon, each hour finds changes in color and shadows. From the early morning's low lights of color, Capitol Peak now gleams in the full light of a midmorning sun, bringing out the full spectrum of the canyon's hues. The foliage also takes on more vivid yellows and greens. A group of cedars, blocking the sunshine, gives a contrast to the bight colors with reddish glowing shadows. Shadows are not dull and gray. All the subtle colors are there if you will look for them. I've used a violet to tie this painting together. Touches are seen throughout the foliage, and Capitol Peak is awash in violets. Look in the sky and you'll see very faint traces of violet.

Note: Paintings 11, 12, 13, 14, and 15 were from a trip to North Cita Canyon with access through the newly acquired Canyon Cita Ranch, an addition to Palo Duro Canyon State Park.

11. North Cita Canyon From the Rim — 2001 — Pastel on Paper 8 x 12

I love to stand on the canyon's rim and look across its expanse. The sun is about to rise. Beautiful plays of light begin to dance around the countryside. In the distance, canyon features dissolve in the haze, while the closest slope in the foreground shows the greatest contrast from the warm sunlit slope to the cool colors in the shadows. We are on Canyon Cita Ranch, an addition to Palo Duro Canyon State Park, and we are heading for the bottom of North Cita Canyon.

12. From Canyon Cita Ranch — 2001 — Pastel on Paper 7 x 11

Here is a small cut in the flat grassland. There are so many of these that they probably don't have names, but they do offer access to the bottom of the canyon, and this one will be our passage to the dry stream bed of North Cita Canyon. Many of the same colorful levels are exposed here as in the main canyon, except on a smaller scale.

13. North Cita Canyon — 2001 — Pastel on Paper 7 x 11

After climbing down hundreds of feet of rock tumble and following a shallow wash, we have reached the bed of North Cita Canyon. It is good to be on flat land again, and the shadows offer some comfort from this August day's heat. My thoughts are not so much about reaching our target, a location near the MacKenzie battle site, but rather about our return trip and the slow arduous journey out of the canyon during the heat of the afternoon.

14. The Cliffs of North Cita Canyon — 2003 — Oil on Board 8 x 10

A walk along the sand bottom of North Cita Canyon brings us to a wall of reds and violets with the very top called Mesquite Mesa. Cross the mesa and the Prairie Dog Town Fork of the Red River is on the other side. So the main portion of Palo Duro Canyon State Park is just a short distance from here, 600 feet up one side and 600 feet down the other. The angles of the separate cliffs in this scene are so important and are needed to lead the viewer through the painting. Start at the sandy bottom of the creek, then follow the gradual slope of the red cliff; now change directions and follow the violet hill. Each angle takes you higher until you reach the top with the final hill's angle. I did not have to create these angles. They were all there naturally.

15. Mackenzie Battle Site — 2002 — Oil on Canvas 12 x 24 — Collection of Joe and Laura Brunson, Dumas, TX

History. It's all around us. You can almost feel it in the air. We are in North Cita Canyon; just around the bend is the confluence with the Prairie Dog Town Fork of the Red River forming a large flat, which is the location of the Indian encampment attacked by Col. Ranald MacKenzie during late September 1874. This was one of the last engagements of the Red River Indian War. The long sloping cliff to the right in the painting is an Indian trail and the route Mackenzie followed to enter the canyon from the rim.

16. Toward Sunday Canyon — 2005 — Pastel on Paper 6 x 20 — Private Collection

It's been a long walk from the road, a good place to stop and do a painting. I've been looking at the openness, a challenge to make something happen. Mostly scrub mesquites and dry grass, but in this case it is the vast distance to the far side that I want. Using an elongated paper will help. Turn nothing into something. Catch the scale to the far side with haze and grayed-back colors. Around the peak to the left are Sunday Canyon and the Lighthouse. It's another two miles that I'll leave for another trip on another day.

17. Sunrise in the Canyon — 1998 — Oil on Canvas 24 x 36

After a night in the canyon, the best part of the day begins. The wind is still down, and slight warmth can be felt as the sun, about to make its appearance for another day sends bursts of color across the sky — never holding the same for more than a few moments. When painting, catch the light and color quickly before they change! Nearby treetops begin to pick up the glow, while shadows continue to hold the coolness left from the vanishing night. Reflected light is already beginning to warm the cliffs, while in the distance the vague features of the Spanish Skirts become visible.

18. Sunset on Sunday Canyon — 2004 — Pastel on Paper 8 x 10 — Collection of Jeanie Martin, Montgomery, TX

A favorite painting location of mine is up Sunday Canyon, a main tributary to the park. The sun is setting but still able to light the tops of the main peaks, while deep shadows have taken over most of the red cliffs. I painted the sunlit areas early and then, when the time was right, established the light-shadow line quickly to work in the colors of the darkening walls.

19. Dawn — 2002 — Pastel on Paper 4 x 8

We are at the same location as the painting "North Cita Canyon From the Rim" (Number 11) before the sun has made its appearance. There is a chill in the air. The best way to paint the feeling you get before sunrise is to limit color. An extreme case is no color. A pastel in black and white with many gray tones in- between creates the feeling I'm looking for and also shows that value is more important than color.

20. Palo Duro Sunset — 2002 — Oil on Canvas 15 x 30 — Collection of Ralph and Eileen Rauton, Amarillo, TX

A late-day thunderstorm crosses the canyon, forcing us to end our day's outing. As we reach the top, the storm breaks, leaving cloud remnants to clash with the setting sun. What a beautiful painting! Shadows have become deep, yet enough sun is left to highlight lines of cottonwoods marching along the creek. The yellows pick up the light, but be careful not to overdo the colors in the Spanish Skirts and other cliffs since they are muted in the final minutes of daylight.

21. Sunset on the Lighthouse — 1997 — Oil on Canvas 18 x 36 — Collection of Jerry and Deanna Thompson, Dumas, TX

Sunsets vie for a painter's attention. Fortunately, they are not a rarity in the Texas Panhandle. Here the Lighthouse is silhouetted against the yellow of the setting sun. The large cliff to the left is deep in shadow and takes on a blood-red appearance, while those to the right are still in the low sunlight and hold their daytime color. I can't spend too much time taking in the beauty but must work quickly to catch the moment.

22. Late Summer Haze — 1983 — Oil on Canvas 16 x 20

    This painting was created at the same location as "Silver and Gold" (number 26) with a slight variation of angle. The two paintings were executed almost 20 years apart. This, the earlier piece produced on a summer afternoon, shows the day's heat settling on the canyon, causing a hazy effect upon the steep walls, while the second painting, worked in the autumn, is crisp as the early morning air. I enjoy the different effects the atmosphere creates on any given day and try to capture this feeling in my paintings.

**23. After a Summer Rain — 2000 — Pastel on Paper 10 x 14
— Collection of Drs. George and Ilona Havasi, Sarasota, FL**

Palo Duro Canyon takes on a different look as late spring, early summer rains swell the streams to their banks, and it's the time of year when everything growing turns green. An easy walk along the creek any other time becomes difficult as undergrowth becomes so dense that it blocks the way. The cool morning quickly changes as the sun moves higher in the sky. Even the shade doesn't overcome the humid air, and I look for a breeze to make its way down the canyon.

24. Hazy Morning — 1983 — Oil on Canvas 14 x 18

Looking through my paintings, you can see how I enjoy haze and fog or anything that will add some mystique to the scene. Here, in a view from the top of the canyon's rim, the early morning haze settles in the valleys. Moving back in space, each range loses more and more clarity, until only the highest peaks can penetrate the film of haze. Study Oriental paintings to see how they mastered this effect a thousand years ago.

25. Late Autumn — Fifth Water Crossing — 2005 — Oil on Canvas 12 x 16

Some scenes just catch your eye and need to be painted. It's almost winter, and the trees are mostly bare, but color is still everywhere in the grasses and hills. I used a very limited palette of four transparent paints and a lot of palette knife work. The paints were transparent red oxide, Indian yellow, Grumbacher's olive green, and ultramarine blue (and of course two whites, flake and titanium). See what you can do with a limited palette.

26. Silver and Gold — 2002 — Oil on Canvas 16 x 20

Complementary colors of violets and yellows caught my eye. This scene is so different from one I did here years ago during the summer. I guess I've changed as much as the season. This crisp autumn morning scene has violet walls of the canyon guiding the way for a wide sweep of a tributary to the Prairie Dog Town Fork of the Red River. The wash is now dry, but silver and gold cottonwoods tell of flowing water during the summer's wet season. Late October and early November can bring the most beautiful weather to Palo Duro Canyon.

27. Red Tamarack — 2000 — Pastel on Paper 10 x 14 — Collection of Drs. George and Ilona Havasi, Sarasota, FL

Cottonwoods are the main trees for autumn color in Palo Duro Canyon, but some of the grasses and tamarack build a competition in color. Look at how some of the reds reflect in the water! This bend in the creek was a nice place to paint, but finding a small flat area to set up created difficulties with the undergrowth, as thick as it was.

**28. Cool Shadows — 1982 — Oil on Canvas
20 x 24 — Collection of the Artist**

On the hottest summer days in Palo Duro Canyon, you can still find cooling relief from the sun in the shadows of trees lining the creeks and washes. This creates a wonderful color combination to paint, the cool blues in the shadows and the warm yellows and golds in the sunlight. Pick a spot in the shade and start painting!

29. The First Water Crossing — 1982 — Oil on Canvas 16 x 20

A slight wind was moving the tree leaves, creating a sparkling mood in the reflecting waters. To make the painting sparkle, small brushwork and rapid value changes were needed. I also used a palette of pure color, avoiding earth paints. Cadmium orange, cadmium red light, and cadmium yellow deep comprise the warm colors while cobalt blue, viridian, and ultramarine violet the cool. My white is soft titanium. Notice these are complementary and analogous colors. It's the palette I use for most of my oil paintings.

30. Hoodoo — 2000 — Pastel on Paper 10 x 16 — Collection of the Carson County, Texas Museum

Standing alone, resisting erosion, the Hoodoo juts from its surroundings, proudly overlooking the sweep of the canyon. The burnt colors of fall have changed the plant life until it blends into the colors of the soil. Cast the reds into the foliage and create a warm painting, but we do need a cool competition. There — a nice turquoise sky will do the job.

31. Red Cliffs Near the First Water Crossing — 2001 — Pastel on Paper 12 x 19 — Collection of Mike & Ann Keck, Amarillo, TX

Brash red cliffs; tall yellow grasses; blue sky reflecting in the water. More color than I normally would put into a painting, but in this case it seems to work. Even more than the color, I think the powerful diagonals caught my attention. The lines of the creek and the grasses lead to the cottonwoods' golden leaves. These meet with a strong diagonal in the opposite direction formed by the top of the cliff.

32. Guardians of Palo Duro — 2003 — Pastel on Paper — 16 x 8

    The sun-baked cliff stands like an army sentinel at attention, guarding the ribbon of water at its feet. The outcropping cliffs become smaller as they wander down the canyon. Color and detail give way to distance and atmosphere. The rules of linear perspective are being applied to this painting. Even the greenery of the canyon floor follows the rules, becoming smaller with duller color and less value change going back into the painting.

33. Creek Trees — 2000 — Oil on Canvas 4 x 5 — Collection of Drs. George and Ilona Havasi, Sarasota, FL

Miniatures are fun to do and don't take much time or paint. I chose this location just after the first water crossing because it has a nice view of the creek in both directions. After I finished this painting I did another one looking the opposite way. Along the dry wash, which angles into the center of the painting, are the remains of an old road that at one time crossed the creek and went back to the Devil's Tombstone. The road has not been passable for years — washed away by floods — although traces of it still can be found.

34. Bottom of the Canyon — 2002 — Oil on Canvas 9 x 12

Changes from shadow to light are interesting in a painting. Not only does this painting have the rapid changes coming toward the water, but also the entire scene is divided with shade to the left and bright sunlight to the right. Small pools formed by the creek act as the dividing line and develop a great counter-move. The downhill play of shadow and light lead to the center of the painting.

**35. Sunlight and Shadows on Red Cliffs — 2002 — Oil on Canvas
11 x 14 — Collection of Tim and Dr. Lois Stickley, Amarillo, TX**

    Two of my favorite times to paint Palo Duro Canyon are early morning and autumn. The sun, low on the horizon, creates long shadows, and these do not have to be dark and grey. Notice the color in the shadows, cool violets and blues, and they are really very light. Where the sun touches the incline, the autumn day comes alive with warm yellows and reds. The contrast is short-lived as the sun marches higher in the sky. Cottonwoods and grasses are also become animated with the sun's warmth, and a slight breeze can stir the leaves into a sparkling dance.

36. Colors in the Canyon — 2004 — Oil on Canvas 24 x 48 — Collection of Monte and Hillary Slatton, Amarillo, TX

When painting a landscape, ask yourself, why do I want to paint this scene? What is it about this location that caught my attention? Then be certain you get that idea into your painting. In this case, the main canyon wall in the background sets a cool backdrop for the warm, red cliffs and golden trees along the creek. All the colors seem to be moving, so I stay away from solid lines and create rapid moving changes in value and temperature.

37. Winter in the Canyon — 2000 — Oil on Canvas 16 x 20

Wind-driven clouds have brought snow to the canyon, which settles into pockets among the shrubs and trees and rocks. The threatening sky is still approaching, bringing the promise of more snow. It's a cold Panhandle day, so make your painting feel that way. Even the usually warm cliffs and grasses have turned cold.

38. After the Snow — 2001 — Oil on Canvas 20 x 24 — Collection of the Rev. Franklin and Barbara Williams, Sekiu, WA

Every season can be beautiful in Palo Duro Canyon. The snow has ended and the sky cleared, leaving the daylight to glisten on the new-fallen snow. No haze today. The cold morning is crisp and clear, with even distant hills appearing more clean-cut than usual. The sun feels good, and the lack of wind feels even better.

39. Winter Fog — 2005 — Oil on Canvas 11 x 14

One of my favorite yet most difficult scenes to paint (at least for me) is fog and low clouds. How do you get the feeling of cold and dampness into your work? Most people think we don't have much fog in the Texas Panhandle due to our dry climate, but it is quite common in the fall and winter months. When painting fog, just the most subtle value changes are needed, and color becomes muted. The bulk of this painting is very cool, but notice the slight tinge of warmth on the nearest cliffs, as if the sun is trying to break through the clouds.

**40. Prickly Pear Cactus — 2002 — Pastel on Paper 5 x 7 —
Collection of Phillip & Christie Nussbaum, Amarillo, TX**

This plant is abundant in Palo Duro Canyon State Park, covering large, flat areas or hanging precariously on the edge of a cliff; any place they may get a little water. Red cliffs and green cactus make a pretty picture, but watch out! This one will punch right through the side of your shoe, and be careful where you sit.

41 Yucca — 2002 — Pastel on Paper 8 x 5 — Collection of Steve & Elizabeth Schafer, Borger, TX

Watch for the first good rains of late spring to set the canyon in bloom, and one of the most prolific flowering plants will be the yucca. Sword-like leaves form a protective barrier for the tall stem of white flowers. This plant is also known as Spanish sword because of its leaves and soap plant because earlier people used its roots as a frothy soap.

42. Grassland — 1997 — Oil on Canvas 24 x 48

To approach Palo Duro Canyon on foot, you walk through fields of wind-blown grass. Gradually there is a slight rise to the ground, and then suddenly the edge of the grass falls away and the gaping canyon lies ahead. One wonders if this is how the early Spanish explorers felt. The poetry and movement in paintings is as important as the quality of the paint. I tried to catch the feeling of the moving grass, first in one direction and then another and another until the entire tabletop of land was in motion. The distant background hills and their stillness form a competition with the moving grass; always competition, whether it is color, direction, value, or movement, it's the building block of a painting.

A DIFFERENT VIEW OF THE CANYON

CONTEMPORARY PAINTINGS

43. Rains Across the Canyon — 1997 — Oil on Canvas 24 x 48

    This is a contemporary, although somewhat visual rendition of the canyon. The feeling of dryness is throughout the painting, but the sky darkens as shafts from the clouds move toward us with the promise of life-giving rain. Direction of brush strokes is very important in this work, apparent in the foreground and flat of the canyon's floor, but also notice the change of direction in the background hills and the vertical alignment in the sky.

44. Storm Clouds and Sun — 2005 — Oil on Canvas 11 x 14

Swirling clouds and then a break of sun. I've seen this so many times while trying to paint, and it's difficult since the light is so changeable. One or the other is fine, but not both at the same time. Well, it doesn't matter today, since this painting is out of my head and I'm sitting in a comfortable studio rather than ducking rain, wind, and insects. The joys of plein air painting.

45. Blue Sky — 2001 — Oil on Canvasette 12 x 16

A small patch of cool blue sky almost outweighs the warm colors throughout the remainder of the painting. The theme of three buttes and mesa appears in this work as well as many of my contemporary pieces.

46. Sunlit Hills and Grass — 2007 — Oil on Canvas 24 x 36

Physically, this painting makes no sense at all, but many contemporary paintings do not. The three mesas and grass are sunlit, but there is no source of sun, since there is darkness beyond the hills, with a layer of grey-white clouds. And this is all topped by a blue-green sky. Each area of the painting is foreign to the next, but it all seems to work as a single entity.

47. Hills and Clouds — 2001 — Oil on Board 30 x 30

Pure color dominates this painting, no grays or earth colors. The three mesas stand out from each other because I used complementary colors, yellows and violets at the bottom, orange and blue in the middle, and green and red at the top. The same applies to the "clouds" forming the diamond at the top of the painting.

48. Clouds and Hills — 2007 — Oil on Canvas 24 x 36

A stylized cloud and the peaks of three mesas are emerging from fog. I like the feeling the complementary colors orange and blue create.

49. Canyon in Red — 2007 — Oil on Canvas 24 x 36

Hot days in Palo Duro Canyon make you feel hot and paint hot. The cool blues in the sky and shadows are not enough to offset the warm reds, oranges, and yellows and yet are necessary to provide a balance of temperature in the painting.

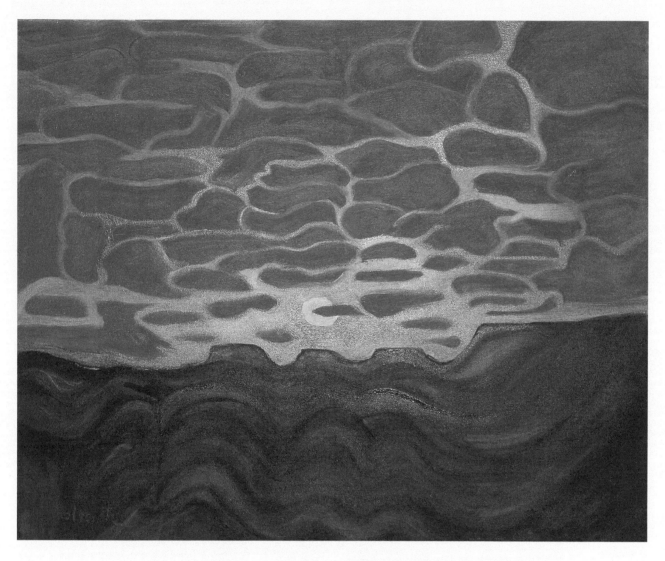

50. Moon Glow — 2004 — Oil on Canvas 16 x 20 – Collection of Tim and Dr. Lois Stickley, Amarillo, TX

A sky full of cloud shapes and the three mesas repeated but with a different treatment. It's night, and the moon is a dull disk casting a glow in the air. Even though it is night, the mesas have warmth to them. This painting was handled with two colors and white, transparent red oxide for the warmth and indigo for the coolness.

51. Cliffs and Trees — 2001 — Pastel on Paper 14 x 10 — Collection of Swift and Gail Lindley, Colleyville, TX

Forget the cliffs and trees and pick out the shapes and colors. These become more important than the objects. The branches of the trees create wonderful shapes, and the colors do too. One edge slowly fades from color to dark while the other ends at an edge of light.

52. Night in the Canyon — 2002 — Oil on Canvas 16 x 20

Night sparks the imagination to see many nonexisting things — mysterious shapes and mysterious colors. This painting has only three transparent colors in a turpentine wash: Indian yellow, Prussian blue, and transparent red oxide. Rather than brushes, rags were used, and what appears white is where the paint was drawn back to bare canvas. I like to experiment with different paints and applications. Either the effect is great or I'll never do that again! I tell students, plan to destroy a canvas occasionally. You're not losing that much, and you'll learn something.

53. Spring in the Canyon — 2007 — Oil on Canvas 24 x 36

The three mesa series continues as spring brings new life to the canyon. Greens replace the reds and oranges of the soil, but never totally. Strong diagonals in this painting walk you back through the canyon to the most distant walls.

54. Blue Moon — 2006 — Oil on Canvas 24 x 36

It's night but not dark. The rising moon creates a vibration in the atmosphere and on the reflection cast on the three mesas. The mood is set with only a few analogous colors, blue and blue-violet, and orange and yellow-orange.

55. Range Fire — 2002 — Oil on Canvas 30 x 40

Fire can be a useful tool when contained, but a mortal enemy when it breaks loose from our control. It leaps with a mind of its own, and the winds are its ally, moving the flame as if in a dance. In this painting the fire is in the canyon and on the grassland above. The colors are as brash as the fire itself, with only the blue sky to cool a small part of the painting. Notice the three butte series is being used but well hidden.

56. Dust Storm — 2002 — Oil on Canvas 30 x 40

In the distance a bank of clouds appear — or so it seems. Coming nearer, the clouds change to a wall, and the wall begins to engulf all in its way. The dust clouds are the same color as the land, making the two hard to tell apart. This storm hides part of the canyon (including the three buttes and a mesa) in soil lifted from the flat land. The violet sky adds a nice contrast to the yellows in the painting.

57. Thunderstorm — 2002 — Oil on Canvas 30 x 20

Rains from thunderstorms are carvers of canyons and producers of life on the plains. Rains from thunderstorms shape the canyon, cutting and moving soil as a craftsman shapes his wares. How many rains have come and gone to bring Palo Duro Canyon to its present form? The never-ending process goes on. At night, lightning flashes bring a fleeting glance of the shapes in the canyon.

58. Emergence 1 — 2002 — Oil on Canvas 24 x 48

This is the first of a three-panel series showing my thoughts on the process used to build Palo Duro Canyon. In the beginning, the land was flat, but small changes began to appear as weathering took place. Breaks from rains formed in this flatness, and the canyon began to evolve. Warm colors dominate this painting, since the climate was warm and dry with only touches of coolness making an appearance.

59. Emergence 2 — 2002 — Oil on Canvas 24 x 48

Time sees the changes taking place. The unpredicted appearance of new shapes emerges in the land. More and more the appearance of the land changes from flat terrain to an undulating landscape, and a movement is forming in the sky. The warmth of Emergence 1 is now giving way to cooler blues, and warm and cool are almost equal. The rains continue to carve.

60. Emergence 3 — 2002 — Oil on Canvas 24 x 48

The canyon lies before us. Shapes familiar today have taken long spans of time to emerge. In this panel, my three buttes and a mesa appear in the foreground, with a series of ridges visible in the distance. The color theme of blue and orange continues with a reversal as cool now dominates. How much time must pass before Emergence 4 can be painted? And how much time must pass before erosion fills the canyon back to the flat plain of Emergence 1? There is no beginning and there is no end.

CHRONOLOGY

1934 Born January 24 in Chicago, Illinois and raised in the west suburb of Elmwood Park.

1951–53 Apprenticed as a tool and die engraver.

1953–57 Served in the U.S. Navy; attended meteorology schools. Stationed at Memphis, Tennessee, Guantanamo Bay, Cuba, and Trinidad, British West Indies.

1958 Married Virginia Isensee and lived in Crystal Lake, Illinois.

1965 Moved to Amarillo, Texas, with Virginia and two daughters, Lois and Gail.

1970 Third daughter, Laura, born in Amarillo.

1970 Studied pastel and oil painting with Ben Konis.

1971 Studied oil painting with Dord Fitz.

1972-73 Attended Amarillo College for drawing, life drawing, and ceramics.

1973-89 Continued studies in oils and pastels with Dord Fitz.

1978 Workshop with Elaine de Kooning.

1980 Studied with Lawrence Calcagno, Taos, New Mexico.

1990 Stone-carving classes with George Bauer.

1993 Life drawing at Amarillo Art Center.

1996 Studied the Impressionist painters in Paris and the Normandy region.

1997 Studied seventeenth- and eighteenth-century paintings and tapestries in England and Scotland.

1998 Oil painting and drawing studies with Emin Abbasov.

1999 Portrait studies in pastels with Steve Napper.

2002 Returned to Paris and Normandy for further studies of the Impressionists.

2004 Ceramic studies Amarillo Art Institute.

2005 Life-drawing studies with Emin Abbasov.

2005 Pastel workshop with Maggie Price.

2006 Conducted a workshop Killarney, Ireland.

2007 Pastel studies with Maggie Price at Juzcar, Spain. 2008 Research and painting in Belize, Guatemala, Honduras, and El Salvador.

Since 1990, taught pastel and oil painting and held numerous workshops in New Mexico, Colorado, Oklahoma, and Texas. Travels since 1996 include studies and painting in twenty-two countries throughout Europe, Asia, Africa, Mexico, Central America, and South America.